“The X-Zyl … igation systems si … (most notably the SR … achines.”

page 38

“With the wrong technique, the X-Zylo™ will fly. With the right technique, it will fly amazingly!”

– *Ryan T., Graduate in Aeronautics and Astronautics at the Massachusetts Institute of Technology (M.I.T.)*

“We were sitting around out office at M.I.T. one afternoon and wondered what made the X-Zylo™ so special. We talked and talked, then finally said “Let’s find out.”

– *Barbara B., Another Graduate in Aeronautics and Astronautics at M.I.T.*

“Look at it go! And now we know why!”

– *Francois L., Yet Another Graduate in Aeronautics and Astronautics at M.I.T.*

Buy an X-Zylo™ and test your own theories!

What Makes the Amazing X-Zylo™ Fly?

Incredible Companion to Your X-Zylo™

David Landon Tarr

Columbia-Capstone

Edition 1.0

X-Zylo™ is a trademark of the William Mark Corporation. The X-Zylo™ can be purchased at stores throughout the world. Buy one and test your own theories.

Published by Columbia-Capstone, Redmond, Washington
www.columbia-capstone.com

Cover design by Brion Sausser, www.bookcreatives.com

ISBN-13: 978-0-9821148-3-4
ISBN-10: 0-9821148-3-4

To my colleagues at M.I.T., students and professors, who apply themselves to solving some of the most interesting and important problems in aeronautics and astronautics.

Contents

Foreword

The X-Zylo toy is amazing and extremely fun to fly. This companion to the toy explains its puzzling physics and fluid mechanics. For anyone curious about how things work, these are hypothesized and proven beyond a reasonable doubt using a series of qualitative experiments coupled with the basic principles of flight. The chapters include modifications to the toy that you can make and try out in flight.

The resulting conclusion is that the primary mechanism responsible for the X-Zylo's lift generation is the gyroscopic effect. When thrown with rapid spin the mass of the annular ring is able to produce sufficient angular momentum to maintain a small angle of attack relative to the X-Zylo's direction of motion. The walls of the X-Zylo's cylindrical body then serve to produce what is known as flat plate lift, redirecting the incoming air downward to provide a reaction force sufficient to keep the X-Zylo aloft. As viscous drag slows the X-Zylo's translational and angular velocity and the downwash is overcome by gravity, the torque produced by the difference of location between the center of gravity and the center of pressure act to turn the angular momentum vector, giving it the characteristic turn to the right or left depending on the direction of rotation. With increasing angle of attack this behavior is exacerbated by travel of the center of pressure aft.

This book and the toy will provide hours of fun, rain or shine. Experiment with some of the concepts by reading about them, modifying your X-Zylo, then taking outside and flying it!

Preface

This book was born out of a typical college student debate.

Four graduate students at M.I.T. were relaxing one Friday afternoon after a long week of classes and exams. We were tired.

I picked up the X-Zylo on my desk and threw it across the office. It didn't take much effort. We started talking about how far it had flown outdoors, much farther than any of us could throw a ball or any type of flying disk.

What made flying the X-Zylo so easy? It didn't look particulary aerodynamic. Describing it in words generally includes something like, "A large can with the ends cut out."

Then I thought, "Why not put a little rigor to our theories?" After all, we were graduate students in aeronautics and astronautics. We should have the basic knowledge to apply to this flight mechanics problem.

It so happened that I was taking a course in Fluid Mechanics and I needed a topic for a term paper. This was it. You now have in your hands an intellectually entertaining attempt at explaining mysteries of the X-Zylo.

Given the origins of the idea for this book, I would like to thank Barbara, Francois, and Ryan for their funny, insightful, debatable, and motivational contributions to the conversation that Friday afternoon at M.I.T. in Cambridge, Massachusetts.

I would also like to thank the editors and publisher at Columbia-Capstone who pushed to get this work completed. Had they not, it would still be sitting on a shelf in draft form.

David Tarr

Sammamish, Washington

March, 2012

What Makes the Amazing X-Zylo™ Fly?

Incredible Companion to Your X-Zylo™

Nomenclature

a angle of attack
ω angular velocity
Ω angular momentum
Λ torque
t airfoil thickness
c average chord length
u_∞ Free-stream velocity in direction of motion
v kinematic viscosity of air
δ boundary layer thickness
γ local vorticity
ń local unit normal vector
u local velocity vector
L lift force
D drag force
d cylinder diameter of X-Zylo
θ transformed chordwise position ($\theta LE = 0, \theta TE = \pi$)
ϑ circumferential location ($\vartheta = 0$ at cylinder's bottom)
m_0 moment about leading edge
I moment of inertia
φ direction of the angular momentum vector

1 Introduction

The X-Zylo is a flying toy that exhibits many unique and largely unexplained flying qualities. Its packaging boasts "NASA experts are amazed," and it has even earned a spot on ESPN which claims it to be the motivation for the fabled gyro-ball that Japanese pitchers are now throwing [1]. Shaped like a cylinder, the X-Zylo is thrown along its axis with rapid spin and a nearly horizontal trajectory, much like a football. Unlike Frisbees, aerobees and other flying toys, the X-Zylo is axisymmetric, leaving the method of lift generation not immediately apparent.

The specific mechanics notwithstanding, the X-Zylo has remarkable stability in the air and if thrown properly can fly in excess of 200 meters, more than two football fields in distance! The goal of this investigation is to explain the key components of the X Zylo's observed flying characteristics using the basic principles of physics and fluid mechanics.

One of the interesting, and most basic, aspects of the X-Zylo's behavior is that no one seems to agree on how it produces lift. Such hypotheses as vortex shedding via the wavy trailing edge, the Magnus effect, and flat-plate lift have been suggested by M.I.T. Aeronautics and Astronautics graduate students. As well, the more subtle dynamics need to be addressed: why it turns sharply near the end of its flight, why sometimes it does not, how the direction of rotation and angular velocity affect its performance (if at all), and the role that the wavy trailing edge plays in its flight (e.g. can it be thrown backwards?).

2 Physical Features

The X-Zylo is a cylindrical tube of plastic 6.1 cm long, as measured to the tip of the trailing edge, and with a diameter of d = 9.7 cm. The plastic body of the X-Zylo is 0.025 cm thick throughout. Inside and abutted to the leading edge is an additional ring of plastic that is 0.104 cm thick and 1.27 cm long in the axial direction. The trailing edge of the device is wavy like a sinusoid with a wavelength of 6.09 cm and an amplitude (depth) of 1.3 cm. With a circumference of 30.47 cm this gives 5 full periods of waves on the trailing edge. From the leading edge (LE) to the trailing edge (TE) the diameter of the cylinder remains constant. Since the walls are also of uniform thickness it follows that the camber of the wall is zero.

Upon weighing it is found that the X-Zylo weighs 22.75 gm with the center of gravity located at an axial distance of 1.3 cm from the leading edge, placing it at 23.8% of the average chord. A diagram of the important features can be found in Table 2.1:

c	d	$C.G.(\%\text{chord})$	m
0.0545 m	0.097 m	23.8	0.02275 kg

Table 2.1. Physical dimensions of the X-Zylo

3 Usage and Flight Behavior

Launching the X-Zylo is similar to the launch of a football. The instructions on the side of the X-Zylo itself read

> Throw Low With Rapid Spin...
>
> Throw Like Football With Ring Forward ...

Experience shows that the two most common difficulties in launching the X-Zylo are:

1. Imparting spin rapid enough to stabilize the X-Zylo
2. Launching the X-Zylo in the intended direction

If both of these criteria are achieved sufficiently, the flight of the X-Zylo is stable and predictable enough to play a game of catch and potentially of ultimate frisbee (but with an X-Zylo). The key, though, is achieving these initial flight conditions.

Upon launching it is apparent that the X-Zylo departs almost immediately from the launch-path to a path along the axis of the cylinder. This is suggestive that the dominant aerodynamic force is that lift produced purely by the angle of attack of the cylinder, and not due to a force arising from the subtle shaping (camber, wavy trailing edge, etc). Intuitively this argument makes sense: wings of an airplane are shaped in a distinctive non-symmetric manner which, once the laws of fluid motion are understood, indicates the direction of the lift force to be generated (even at $\alpha = 0$). An axisymmetric object such as the X-Zylo has no geometric features that could force the airflow in a consistently favorable direction. Even if there were a force generated by the loading and unloading of the wavy trailing edge - as one student suggested - there could be no net force generated since the object is uniform about its circumference. From this we can conclude that there cannot be a purely geometric lift force generated by this axisymmetric X-Zylo body.

If the initial launch conditions are met and the X-Zylo is given a good trajectory the flight is relatively long. During the flight it appears that the X-Zylo is producing lift at a zero angle of attack ($\alpha = 0$). This fact has led some to conclude that the lift is generated through means not related to angle of attack. It is the opinion of the author, and his goal to prove, that the angle of attack required for flight, α_{req}, is simply imperceptibly small given the X-Zylo's weight. Toward the end of such a flight an additional peculiar behavior can be witnessed. As the drag slows the forward velocity and the X-Zylo begins descending, it yaws to the left! The degree of yaw is quite dramatic, easily turning the axis of the cylinder nearly 90° from the original orientation. Suggestions for the cause of this behavior have centered around Magnus forces created by the spinning cylinder to gyroscopic precession due to a yet to be determined torque acting on the cylinder.

4 Explanation of Flight Behavior

From observations made while using the X-Zylo and an off-the-cuff engineering intuition, it is hypothesized that most of the flight behaviors that the X-Zylo exhibits can be explained by the combination of flat-plate aerodynamics and gyroscopic principles. The following sections are aimed at justifying this idea through application of known methods.

Thin Airfoil Theory – The paramount question is that of lift production. Since the X-Zylo has no camber to speak of, can it produce enough lift to support itself while kept at a small, but constant, angle of attack? The method of maintaining this angle of attack is not of importance for now. To attack this problem we need to simplify it to be able to use the inviscid tools from Fluid Mechanics and other topics.

Local Thickness – To assume that the cross-section of the X-Zylo behaves like an infinitely thin flat plate, it must be justified that the upper and lower surface profiles are similar enough to that of a truly thin flat plate, $t/c = \infty$. When examining physical features, we determined that the cross section is of constant thickness (with exception of the leading edge weighting ring), and zero camber. If the thickness to chord ratio is also very small, the flat-plate assumption will be justified. Using the measured dimensions of the X-Zylo we see that this is the case:

$$t/c = \frac{0.025}{5.45} = 4.58 \times 10^{-3} \tag{4.1}$$

Reynolds Number – To weigh the relative importance of the inertial forces to the viscous forces of the X-Zylo being launched in air, we calculate the Reynolds number. If we estimate that the X-Zylo is thrown with a velocity of $U_x = 10\ m/s$, then the Reynolds number is:

$$\mathbf{Re} = \frac{U_x c}{\nu} \sim \frac{10 * 0.0545}{1.46 * 10^{-5}} = 37328.8 \tag{4.2}$$

Therefore, we can assume that on the scale of the system in question, the effects of viscosity are small compared to inertial effects. Although the Reynolds number is very large, near the surface of the airfoil viscosity is important. As we discussed in relation to the circulation about a cylinder, it is the generation of vorticity through viscosity that allows for the generation of lift in the first place. Since viscosity is the key enabler to lift and cannot be eliminated altogether, two more assumptions will allow us to neglect it in the calculations while qualitatively accounting for its effects.

The Kutta Condition – The first assumption directed towards dealing with viscosity is the imposition of the so-called Kutta condition at the trailing edge. In a potential flow there is no manner in which vorticity can be imparted due to geometry only. This means that for an airfoil at an angle of attack there is a stagnation point on the lower surface near the leading edge and on the upper surface near the trailing edge resulting in zero net circulation as shown in Figure 4.1.

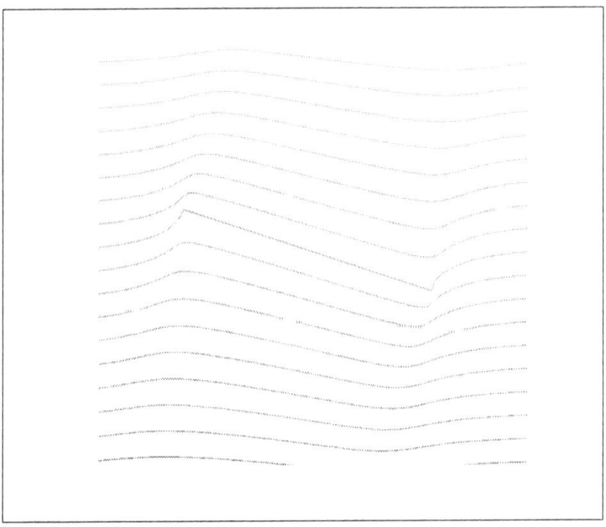

Figure 4.1. flat plate at $\alpha = 15^{\circ}$ with zero circulation.

Since the body's surface must also represent streamlines, via the no-flux boundary condition, there must be an infinite pressure gradient at the leading and trailing edges to produce the discontinuous change in velocity that forces the flow around those corners. In a high Reynolds number viscous flow these infinite velocity gradients are met with a rise in magnitude of the viscous forces locally. If we examine the 2-dimensional Navier-Stokes equations for the unique case of a flat-plate trailing edge, as $\partial r \longrightarrow 0$ we have:

$$\left(\lim_{\partial r \to 0}\right) \quad \rho \frac{u_\theta^2}{r} = -\frac{\partial p}{\partial r} + \frac{2\mu}{r^2}\frac{\partial^2 u_\theta}{\partial \theta^2} \tag{4.3}$$

For this case we have recognized that as the flow make the turn at r = const, the quantities of interest are those which relate the rate of change of the $u\theta$ with changes in θ itself. When the momentum equation is simplified in this manner it can be seen that when r is small (and ultimately tending towards zero) tremendous viscous forces are generated due to the *1/r2* term, and an equally large pressure gradient is required to turn the flow.

Quantitatively the effect of these forces can be calculated numerically using a local-scaling of the thin shear layer equations in terms of the stream function variables, ψ (stream-function), *u* (velocity), and τ (shear). Making this transformation (x, y) → (ξ, η), where $\eta = y/\sqrt{\nu\xi/u_{edge}}$, and substituting (ψ, *u*, τ) → (*F*, *U*, *S*) gives the characteristic Falkner-Skan 3rd order differential equation:

$$F''' + \frac{\beta_u + 1}{2} - FF'' + \beta_u\left(1 - F'^2\right) = 0 \qquad \text{where } F' = \frac{\partial^2 F}{\partial \eta^2} \tag{4.4}$$

β_u is essentially a scaled pressure gradient related to the term $u_x \frac{\partial u_x}{\partial x}$, which drives the flow in the thin shear layer equations. We now can rewrite equation 4.4 as a reduced order system of ODEs with boundary conditions as follows:

$$F' - U = 0 \tag{4.5}$$

$$U' - S = 0 \tag{4.6}$$

$$S' + \frac{\beta_u + 1}{2} FS + \beta_u\left(1 - U^2\right) \tag{4.7}$$

$$\text{B.C.s}: \quad F(0) = 0 \quad U(0) = 0 \quad U(\eta_{edge}) = 1 \tag{4.8}$$

When this system is solved numerically, much information about laminar boundary layer behavior can be gained. The most important result for our purposes is that for a flat plate the edge velocity of the flow can be approximated as a power-law,

$$u_{edge} \sim x^a \qquad \text{where } a = \frac{\beta_u}{2 - \beta_u} \tag{4.9}$$

When the term *a* turns negative, this implies the flow outside of the boundary layer is slowing due to an adverse pressure gradient. As this occurs, solutions to the Falkner-Skan equations show the velocity profile within the boundary layer begins to inflect, shown in Figure 4.2. At a value of $\beta_u = -.1988$, the flow near the wall reaches zero forward velocity and the flow has separated. Equation 4.9 defined how the thickness of the boundary layer scales with edge velocity. Using this and the new power law for the edge velocity we see that when $\beta_u = 1$, the edge velocity scales with x such that

$$\delta_x = \frac{x}{\sqrt{u_{edge} x / \nu}} = \text{constant}$$

This situation corresponds to that of stagnation point flow, when the free-stream direction is oriented perpendicular to the wall.

When $\beta_u = 0$, the edge velocity is independent of x such that:

$$\delta_x = \frac{x}{\sqrt{u_{edge} x / \nu}} \sim \sqrt{x}$$

This solution is consistent with the classical case of Blasius flow, where the free-stream is oriented parallel to the wall.

Accordingly, when $\beta_u = -.1988$, the case of separating flow, the corresponding wall angle is easily found to be -17.9 degrees.

This result can be used to address the flow behavior in the regions of the leading and trailing edge that could not be satisfactorily addressed by potentially flow theory. For the X-Zylo the thickness of the boundary layer is on the order of the thickness of the sharp trailing edge. In fact, for the X-Zylo, these values are nearly identical: $t/\delta = 0.88$! Because the viscous length-scale is of the same order as the physical length-scale, we must consider the effects of viscous and inertial forces to be of the same magnitude as we round the edges of the airfoil.

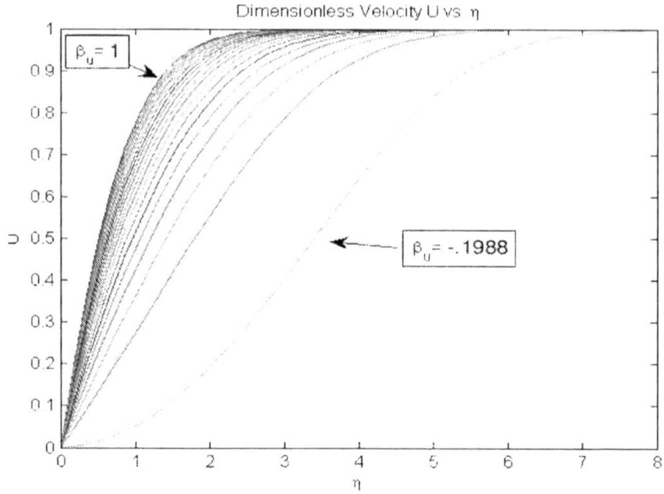

Figure 4.2. An example of numerical velocity profile solutions to the Falkner-Skan equations for a range of β_u values.

With this in mind we now apply what we observed in the Falkner-Skan analysis to the case of flow about a thin-flat plate. As flow travels from the leading edge stagnation point (the point of highest static pressure), it experiences a favorable pressure gradient up to and through its acceleration around the leading edge. It is not until it reaches the upper surface (which is tilted away from the free-stream) that it encounters an adverse pressure gradient. Because the static pressure right at the leading edge is very small, the return to a pressure on the order of the dynamic head ($\frac{1}{2}\rho u^2$) brings a very adverse pressure gradient. It is easily shown that this pressure gradient greatly exceeds the pressure gradient experienced on the flat plate inclined at −17.9 degrees away from the flow, thus separation occurs in nearly every case. This region of stalled flow (or *separation bubble*) is aided in reattachment by momentum transfer from the incoming free-stream air. This phenomenon is of particular importance on strike-fighter aircraft, which utilize thin airfoils with sharp leading edges for supersonic performance. On these airfoils separation bubbles form at the leading edge during low speed flight at any angle of attack and may or may not reattach to the surface of the airfoil. To help alleviate this problem and reduce the stall speed, most fighter-aircraft have leading edge flaps to better align the incoming flow with the local chord [2]. Typical subsonic aircraft work around this problem with rounded leading edges.

In the case of the X-Zylo, as long as α is small, then β_u coupled with free-stream momentum diffusion will be sufficiently large to keep the flow attached over the remainder of the upper surface.

Conversely, when the flow on the lower surface attempts to make the turn around the trailing edge, the flow separates in a dramatic fashion. Unlike the leading edge, when the trailing edge flow initially separates following the corner, the momentum diffusion from the free-stream perpetuates separation. For this reason the flow separates completely right at the sharp trailing edge. With the flow on the upper surface the wing experiencing a relatively minor adverse pressure gradient (again, at small values of α), it is able to continue to the trailing edge fully attached. The combination of these effects moves the second stagnation point to the sharp trailing edge, satisfying the condition that the local velocity vectors on the surface are parallel to the surface itself, and providing the induced circulation

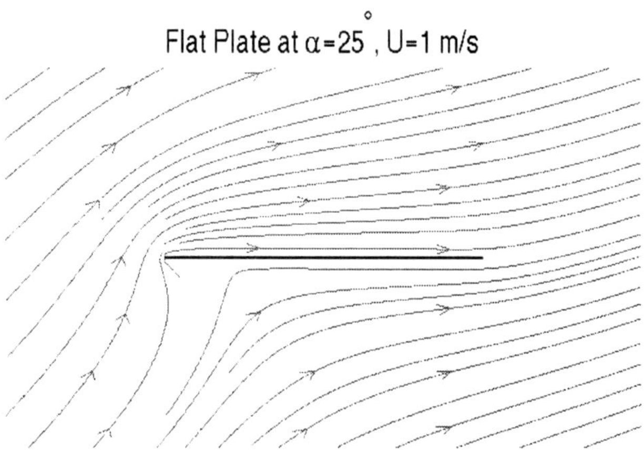

Figure 4.3. Flat plate airfoil producing lift with Kutta condition applied.

necessary for lift (see Figure 4.3). Mathematically this means that the local vorticity at the sharp trailing edge is zero

Although this analysis is extremely helpful in providing the physical basis for the Kutta condition, it is not practical to solve the problem of the X-Zylo with this level of depth. Instead we simply take the result found above, the Kutta condition, and apply it as a boundary condition to the inviscid problem. Such methods have been found to agree satisfactorily with experiment [3].

Angle of Attack – The second additional assumption needed to neglect viscosity in this analysis is that the angle of attack, α is small. As we saw in the analysis leading to the usage of the Kutta condition, small angles of attack ensure: By requiring that α is small, we ensure:

- Diffusion of momentum from the free-stream minimizes the size of the separation bubble
- Small adverse pressure gradients along the chord to:
    1. keep boundary layers attached to the airfoil
    2. keep boundary layers thin compared to the chord length, c

As long as this is true, the streamlines within the boundary layer run parallel to the free-stream flow and the static pressures between the two are equal. Since the static pressure at the surface equals that of the free-stream, there is no appreciable effect of the boundary layer on the net force felt by the airfoil perpendicular to the airfoil surface. As well, keeping a small keeps the separation inherent to flat plates (from The Kutta Condition section) to a minimum. To ensure that our assumption of a thin boundary layer is accurate we must calculate its thickness relative to the length scale of the problem, c. It can be shown that the thickness of the boundary layer scales with the inverse square root of the local Reynolds number [4]. Comparing this to the chord length we have:

$$\delta = \frac{c}{\sqrt{\mathbf{Re}}} \Rightarrow \frac{\delta}{c} = \frac{1}{\sqrt{\mathbf{Re}}} = 5.17 \times 10^{-3} \qquad (4.10)$$

Indeed, the relative thickness of the boundary layer is expected to be very small.

To summarize, the assumptions used in the analysis are:

- $t/c \ll 1$
- $\mathrm{Re} \gg 1$
- α is small such that $\cos\alpha \approx 1$
- $\gamma_{TE} = 0$ (Kutta condition)

With these assumptions under our belt, we are now ready to proceed with the analysis. The following analysis leading up to equation 4.4 is taken from [2]. Lift is mathematically generated by circulation about a body in the flow. To generate circulation we define a profile of vortex filaments of infinitesimal strength, $\gamma(s)$, along the camber line of our airfoil.

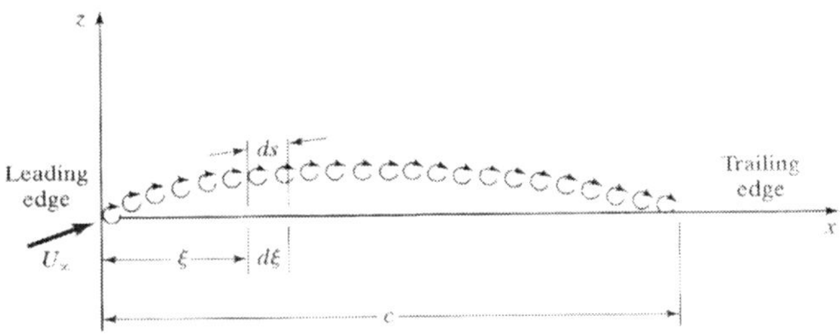

Figure 4.4. Representation of the mean camber line of an airfoil by a vortex sheet whose filaments are of strength γ(s) ([2]).

The boundary conditions on equation 4.9 are: 1.) $u \perp ds = 0$ (no flux), and 2.) $\gamma(c) = 0$ (Kutta condition). Since the local vorticity at any point contributes to the camber-normal flow velocity at a point, P, by:

$$\Gamma = \int_0^c \gamma(s)ds \qquad (4.11)$$

(Kutta condition). Since the local vorticity at any point contributes to the camber-normal flow velocity at a point, P, by:

$$du_{s,\perp} = \frac{\gamma\, ds \cos\delta_3}{2\pi r} \tag{4.12}$$

where the relative angles between the influencing vortex and the point in question are defined as in Figure 4.5, below.

To accurately capture the combined effect of each infinitesimal vortex, we must integrate the effect over the entire airfoil. We will define $\xi = x$, but use the variable ξ in reference to the chordwise location of a particular vortex filament, γds, and use x when referring to a point, P, affected by that vortex element.

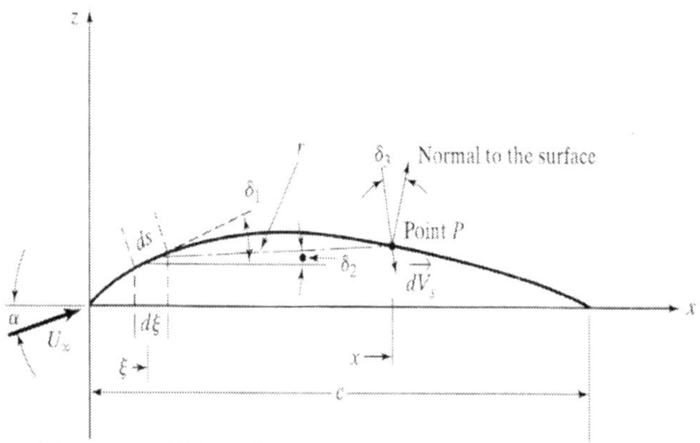

Figure 4.5. Thin airfoil geometry parameters,(from [2]).

Recognizing that $r = \frac{x-\xi}{\cos\delta_2}$ and $ds = \frac{d\xi}{\cos\delta_1}$ the vortex-induced velocity at any point P is:

$$u_{s,\perp}(x) = \frac{-1}{2\pi}\int_0^c \frac{\gamma(\xi)\cos\delta_2\cos\delta_3 d\xi}{(x-\xi)\cos\delta_1} \tag{4.13}$$

As well, there is a component of the free-stream velocity normal to the airfoil's local camber that can be expressed as,

$$u_{\infty,\perp}(x) = u_\infty \sin\left(\alpha - \tan^{-1}\frac{dz}{dx}\right) \tag{4.14}$$

where dz/dx is the local surface slope relative to the free-stream. Since the net flow normal to the surface must be zero everywhere we can say $u_{\infty,\perp}(x) + u_{s,\perp}(x) = 0$. For our flat-plate $\delta 1 = \delta 2 = \delta 3 = dz/dx = 0$, and the expression reduces to:

$$\frac{1}{2\pi}\int_0^c \frac{\gamma(\xi)}{(x-\xi)} d\xi = u_\infty \alpha \tag{4.15}$$

making the coordinate transformation $\xi = (c/2)(1 - \cos\theta)$, and $x = (c/2)(1 - \cos\theta_0)$ gives:

$$\frac{1}{2\pi}\int_0^\pi \frac{\gamma(\theta)\sin\theta}{(\cos\theta - \cos\theta_0)} d\theta = u_\infty \alpha \tag{4.16}$$

Bertin, [2], has found the vortex distribution which satisfies this no-flux equation as well as the Kutta-condition to be:

$$\gamma(\theta) = 2\alpha u_\infty \frac{1+\cos\theta}{\sin\theta} \tag{4.17}$$

By applying this distribution to the the Kutta-Joukowski theorem the expression for the lift force per unit span can be found to be:

$$L = \int_0^c \rho_\infty u_\infty \gamma(\xi) d\xi \quad \rightarrow$$

$$L = \rho_\infty u_\infty^2 \alpha c \int_0^c (1+\cos\theta)\, d\theta \quad \rightarrow \quad L = \pi \rho u_\infty^2 \alpha c \tag{4.18}$$

However, since our airfoil is not an infinite wing, but a cylinder, this lift force must be integrated around the annulus, where the relative angle of attack is not constant. If we define the lowest point on the cylinder to be $\vartheta = 0$ then the relative angle of attack at any circumferential position, ϑ, should be:

$$\alpha_{effective} = \alpha |\cos\vartheta| \tag{4.19}$$

Introducing this expression into the formula for lift and integrating around the circumference of the cylinder we find:

$$L_{cyl} = 2\pi\rho u_\infty^2 \alpha c \int_{\frac{-\pi}{2}}^{\frac{\pi}{2}} \cos\vartheta \frac{d}{2} d\vartheta \Rightarrow L_{cyl} = 2\pi\rho u_\infty^2 c d\alpha \tag{4.20}$$

For the condition of straight and level flight the lift force must be of equal magnitude to the weight of the X-Zylo, which we determined to be 22.75

grams. For an average throw of velocity $u_\infty = 15\ m/s$, in air (ρ =1.29 kg/m^3), the angle of attack required to produce this lift is:

$$\alpha_{req} = \frac{m\,g}{2\pi\,\rho u_\infty^2\,cd} = 1.32^\circ \tag{4.21}$$

This result aligns well with experience using X-Zylo: the required angle of attack is imperceptibly different than zero to casual human observers.

It is important to note that as the circumferential position varies and the effective angle of attack diminishes, a not immediately insignificant span-wise flow is induced, simulating a swept wing section. If we decompose the flow into its normal (chordwise, $\perp$) and tangential (spanwise, $\|$) components, the tangential component flows spanwise without producing a disturbance.

$$u_\perp = u\cos\alpha \quad u_\| = u\sin\theta\sin\alpha \tag{4.22}$$

A more rigorous look reveals that when the governing equations are written for an infinitely long, swept wing with the coordinates shifted so that z is tangent to the span and x perpendicular to the span, they reduce to, according to [4],

$$\frac{\partial u}{\partial x} + \frac{\partial v}{\partial y} = 0 \tag{4.23}$$

$$u\frac{\partial u}{\partial x} + v\frac{\partial u}{\partial y} = u_\infty \frac{\partial u_\infty}{\partial x} + \frac{\partial}{\partial y}(\mu \frac{\partial u}{\partial y} - \rho \overline{u'v'}) \tag{4.24}$$

where the quantity $\rho \overline{u'v'}$ represents a Reynolds stress. These are the same relations as for strictly two dimensional flows, which means that analysis normal to the span is all that need be considered and that cross flow has no affect on the flight of the X-Zylo.[1]

[1]This holds true both for laminar boundary layers and for high Reynolds number flows. For low speed turbulent flows, the perturbation velocity v' is no longer insignificant and the solution has a slight dependence on 3D effects [4].

5 Detailed Flight Mechanics

Stability – As with any aerodynamic device, the X-Zylo must have a mechanism by which it can dampen small perturbations and maintain steady flight. Conventional aircraft accomplish this by tailoring the relative locations of the center of gravity and the lifting surfaces. Consisting of only a single element, it is proposed that the X-Zylo maintains stability by a combination of aerodynamic tailoring and the gyroscopic effect.

Since all of the dominant forces act in the vertical plane we need only consider the stability there. To find the pitching moment about the leading edge due to the pressures acting over the chord we integrate the equation:

$$m_0 = -\frac{1}{2}\rho u_\infty^2 \alpha c^2 \int_0^c \rho_\infty u_\infty \gamma(\xi)\,\xi\,d\xi \rightarrow m_0 =$$
$$-\rho_\infty u_\infty^2 \alpha c^2 \int_{-\pi/2}^{\pi/2} \int_0^\pi \left(1 - \cos^2\theta\right)(\cos\vartheta)\frac{d}{2}\,d\theta\,d\vartheta \tag{4.25}$$

where the integration of the ϑ terms account for the cylindrical profile. The result of this integration is:

$$m_0 = -\frac{\pi}{2}\rho_\infty u_\infty^2 c^2 d\alpha \tag{4.26}$$

For simplicity, we can think of the lift force as acting at one point on the body, the location of which must satisfy the moment about the leading edge which we just found. This point of concentrated lift is called the *center of pressure* and can be found by equating 4.26 with 4.20 multiplied by the location representing the center of pressure, x_{cp}.

$$m_0 = -L_{cyl} x_{cp} \rightarrow -\frac{\pi}{2}\rho_\infty u_\infty^2 c^2 d\alpha = 2\pi \rho u_\infty^2 c d\alpha \, x_{cp} \quad (4.27)$$

$$x_{cp} = \frac{c}{4} \quad (4.28)$$

This remarkable result reveals that the center of pressure for a flat plate occurs neatly at a distance of 1/4 chord from the leading edge. It can also be shown that for a cambered airfoil, the aerodynamic center also occurs at 1/4 chord. According to Bertin, [2], even when the effects of viscosity are accounted for, the aerodynamic center falls between 0.23 and 0.27 of the chord. In section 2 we found that the center of gravity for the X-Zylo occurred at 0.238 chord. Since the relative positions of the lift force and the gravitational force are so close together they are likely to produce very small torques about the leading edge. The result is a natural stability, and one very important feature that produces smooth and consistent flight characteristics.

The other major stabilizing force designed into the X-Zylo is that of the gyroscopic effect. By imparting rapid spin at launch, it is possible to produce an angular momentum vector so large that any perturbing torque (produced by a cross-wind gusts or turbulence) is not strong enough to force a significant change in attitude. This same gyroscopic property has been put to work in the inertial navigation systems of naval warships, spy planes (most notably the SR-71) and other high performance machines with great success. The magnitude of the annulus' angular momentum is defined by:

$$|\vec{\Omega}| = I\omega \quad \text{where} \quad I = m\left(\frac{d}{2}\right)^2 \tag{4.29}$$

We will see the importance of the value of Ω when we consider a perturbing torque in the section on Gyroscopic Yaw.

Induced Drag – Since the inviscid model used to deduce the lift force on the X-Zylo finds zero net-drag (D'Alembert's paradox), we must use another method to estimate drag. To calculate the induced drag we will use two known approaches and take the conservative (larger) of the two values as the accepted value. The first approach will be the purely trigonometric approach. Since pressure forces act normal to surfaces, the drag force can be calculated as the corresponding force in the direction opposite to u_∞ that satisfies the lift constraint for the given angle of attack, α.

$$D_{induced} = L_{cyl} \tan \alpha \tag{4.30}$$

The second approach is taken from Demasi, [5], who authored an excellent mathematically founded paper on the theoretical efficiencies of closed and C-wing systems. His approach to lift calculation is very similar to the one taken in this book with the exceptions that he uses doublets and begins his calculations assuming an arbitrary annular doublet distribution whereas here the approach of a straight line vortex distribution is taken and subsequently integrated around the annulus as α varies with θ. According to Demasi's doublet approach, the minimum value for induced drag for the annular wing can again be found as a function of the prescribed lift:

$$D_{induced} = \frac{1}{\pi \rho} \left(\frac{L_{cyl}}{u_\infty d} \right)^2 \tag{4.31}$$

Through substitution it can be seen that the ratio of Demasi's drag force to the trigonometric drag force is:

$$\frac{2c\alpha}{d\tan\alpha} \rightarrow \frac{2(5.45)}{(9.7)} \sim 1.12 \qquad (4.32)$$

since $\alpha/\tan\alpha$ is very nearly 1 for all small values of α. Therefore we will use Demasi's expression for induced drag since it is slightly larger and therefore a more conservative estimate.

Viscous drag – For calculation of viscous drag we will assume at all times an attached, laminar boundary layer and a constant edge velocity equal to the free-stream value, u_∞ (appropriate for $\alpha \ll 1$)[2]. We can estimate the shear force applied by the laminar boundary layer from the Falkner-Skan equations outlined in the section on the Kutta Condition and calculate the net force due to the viscous generation of vorticity at the surface of the airfoil. The total drag per unit span is simply the integral of the wall-shear over the chord:

---

[2] A simulation of the device at standard launch conditions has been executed using a code written by the author based on interacting boundary layer theory (IBLT), a quasi-one dimensional, viscous, variable u_{edge}, and turbulence capable model. The result showed no transition to turbulent flow, justifying these simplifying assumptions.

$$D_{viscous} = \pi d \left[\int_0^c \tau_{upper} ds + \int_0^c \tau_{lower} ds \right] \quad \text{where} \quad \tau = \mu \frac{\partial u}{\partial y} |_{y=0} \tag{4.33}$$

Since the airfoil is at a small angle of attack the wall shear on each side will not be the same. The factor of πd accounts for the total surface area of the cylinder's annulus. The expression for $\tau|_{y=0}$ is regularly found in texts for a flat plate at zero angle of attack, but for our purposes we would like to allow the drag to change as the angle of attack changes (within the $\alpha \ll 1$ constraint) to reflect changes in conditions that would affect the trajectory or stability. For illustrative purposes, the viscous drag is calculated for our standard launch of velocity $u_{inf} = 15 m/s$ and $\alpha = 1.32$ degrees. From the transformed Falkner-Skan equations we have:

$$\tau = \mu \frac{\partial u}{\partial y} = \frac{\partial u}{\partial \eta} \frac{\partial \eta}{\partial y} = \frac{\mu u_e^{3/2}}{\nu^{1/2} s^{1/2}} S \tag{4.34}$$

The value of the velocity gradient at the wall, S, is found numerically to be $S_{upper} = 0.31719$ and $S_{lower} = 0.34657$ for the standard case. Integrating this known expression for the viscous drag along the chord and around the annulus gives:

$$D_{viscous} = \frac{\pi \mu d\, u_e^{3/2} c^{1/2}}{\nu^{1/2}} (S_{upper} + S_{lower}) \tag{4.35}$$

Combining this with the induced drag, we have a complete zeroth-order representation of the drag forces experienced by the X-Zylo in the axis of velocity. For the standard case the drag on the X-Zylo at the moment just after launch is:

$$D_{total} = 0.025639 \quad N \tag{4.36}$$

using the properties of air at STP from [6]. It is surprising that the lift to drag ratio is less than 1 for these conditions. When examined closely, though, the skin-friction drag accounts for a force 400 times greater than the induced drag. Thus the low L/D ratio is a consequence of the minimal lift required by the X-Zylo to fly, not that the lifting efficiency is so poor as the traditional interpretation of the L/D ratio insinuates.

The second dimension of importance in terms of viscosity is in that of rotation. As the X-Zylo spins, viscous forces reduce the angular momentum of the device by imparting momentum to the air around it. Eventually the angular momentum deteriorates to a point where it is no longer large enough to stabilize the orientation against exterior forces. It is important to develop a simple model for this effect as well to better understand and predict X-Zylo behavior.

Our initial approach is to evaluate the case of rotating concentric cylinders using the Navier-Stokes equations. As was shown in the equations surrounding 4.20, we can consider the forward flow problem decoupled from the spanwise problem in the high Reynolds number limit. Although solutions exist for the viscous torque about a cylinder, the underlying assumption is sometimes that of Re→ 0, which is certainly not the case here. Therefore we will simply solve a new, yet simple, problem posed by two rotating, concentric cylinders with radii R_1, R_2 and angular velocities ω_1,ω_2. Additionally we will assume that the flow is steady and fully developed. The governing momentum equations for this scenario are:

$$-\frac{u_\theta^2}{r} = -\frac{1}{\rho}\frac{dp}{dr} \tag{4.37}$$

$$0 = \mu\frac{d}{dr}\left[\frac{1}{r}\frac{d}{dr}(r\,u_\theta)\right] - \frac{u_\theta}{r^2} \tag{4.38}$$

integrating equation 4.38 twice yields:

$$u_\theta = \mathrm{A}r + \frac{\mathrm{B}}{r} \tag{4.39}$$

with boundary conditions $u_\theta(R_1) = R_1\omega_1$ and $u_\theta(R_2) = R_2\omega_2$. When this is done we find:

$$A = \frac{\omega_2 R_2^2 - \omega_1 R_1^2}{R_2^2 - R_1^2} \quad \text{and,} \quad B = \frac{(\omega_1 - \omega_2) R_1^2 R_2^2}{R_2^2 - R_1^2} \tag{4.40}$$

Substituting these constants into the expression found for u_θ,

$$u_\theta = \frac{1}{1 - (R_1/R_2)^2} \left\{ \left[\omega_2 - \omega_1 \frac{R_1^2}{R_2^2} \right] r + \frac{R_1^2}{r} (\omega_1 - \omega_2) \right\} \tag{4.41}$$

In the case of the X-Zylo, there is no outer cylinder, so we evaluate this solution in the limit that $R_2 \rightarrow \infty$ and $\omega_2 = 0$. Then the simplified expression is:

$$u_\theta = \frac{R_1^2}{r} \omega_1 \tag{4.42}$$

The shear stress for the X-Zylo in cylindrical coordinates is:

$$\tau_{r,\theta} = \mu \left[r \frac{\partial}{\partial r} \left(\frac{u_\theta}{r} \right) + \frac{1}{r} \frac{\partial u_r}{\partial \theta} \right] \rightarrow$$

$$\tau_{r,\theta} = \frac{-2\mu \, \omega_1 R_1^2}{r^2} \text{ which, at } r = R_1 \text{ is } \tau_{R1,\theta} = -2\mu\omega_1 \tag{4.43}$$

To find the torque applied on the X-Zylo by the viscous shear stress on its exterior surface we simply integrate the shear stress at the wall, R_1, around the annulus and multiply by the chord length over which it acts, c, and the moment arm, $d/2$.

$$\Lambda_{ext} = \frac{d}{2} c \int_0^{2\pi} \tau_{R1,\theta} \frac{d}{2} d\theta = -\mu c d^2 \omega_1 \tag{4.44}$$

To find the torque applied on the interior surface we simply rewrite the boundary conditions to be $u_\theta(0)=0$ and $u_\theta(R2) = R_2\omega_2$ (Note that the subscript, *2*, now represents the interior surface of the X-Zylo.) The solution is then easily found to be:

$$u_\theta = \omega_2 r \tag{4.45}$$

It can be seen from this solution that the wall shear on the interior is then zero for this case. The assumption that the rotational flow is fully developed and steady in time led to the result of solid-body rotation for the interior. In reality the flow is not fully developed nor is it steady. Although these assumptions were necessary to find an analytical solution, our intuitive knowledge of fluid mechanics cannot justify that the result is comparable to reality. Therefore another solution must be used.

Since the ratio of boundary layer thickness to local radius of curvature in the ϑ-direction is much less than 1, $\delta/(d/2) = 0.0058 \ll 1$, it is possible to use the same Falkner-Skan numerical solutions for the shear stress at the wall and integrate around the cylinder. Using the flat-plate value of the non-dimensional shear, $S_{\beta_u=0} = 0.33206$, an integration can be carried out to find the torque, Λ:

$$\Lambda = 2\frac{d}{2}c\int_0^{2\pi} \tau_{r,\vartheta}\frac{d}{2}d\vartheta = \frac{2\pi^{3/2}\mu d^{5/2}\, u_{e,\theta}^{3/2}}{2\nu^{1/2}} S_{\beta_u=0} \tag{4.46}$$

where the factor of 2 accounts for the interior and exterior surfaces, $d/2$ represents the moment arm, u_e is evaluated as the velocity of $u_{r=R1+\delta,\theta}$ from equation 4.3.3, and πd replaces c. When numbers are applied, the viscous torque applied for the standard launch ($\omega = 94\ rad/s$) is:

$$\Lambda = 3.67x10^{-4}\ N\cdot m \tag{4.47}$$

A very small torque indeed. However, integrated over the duration of flight the reduction in angular momentum is significant. For a ~10 second flight if the rate of change in the angular momentum is equal to the torque. If we approximate these changes as being linear, for illustrative purposes, we see that:

$$\Delta\Omega = \Delta t\,\Lambda = (10)(3.67x10^{-4}) = 0.00367\ N\cdot m\cdot s \tag{4.48}$$

Since Ω at launch is 0.00503 $N{\cdot}m{\cdot}s$, this is a reduction of 73% of the angular momentum! Although the true relationship is not linear, this simple calculation shows that the methodology is capable of describing the motion well, should a more comprehensive and time-accurate model be developed in the future.

Gyroscopic Yaw – During times of high angle of attack, the X-Zylo is seen to yaw significantly to the left if a clockwise rotation is imparted at launch. Through experimentation, it is found that a counter-clockwise rotation induces a yaw to the right. It is hypothesized that the inviscid influence of the lower surface on the upper surface causes an unequal lift production between the upper and lower surfaces of the X-Zylo. Although the angles of attack experienced are assumed to be insufficient to produce flow separation, they are enough to produce a significant torque on the device. Using a potential flow model, the effect of the coupled-airfoil system can be quantified. The solutions have been examined in detail by several people, including [7], [8] and [9]. The analytical solutions they have found for this problem are beyond the scope of this book, but the results can be applied to explain the observed behaviors.

Assuming that a clockwise rotation is imparted to the X-Zylo at the time of launch (as viewed from behind), the X-Zylo will begin a gentle left-hand yaw that accelerates as the flight speed and angular velocity decrease. The only physical mechanism that readily explains this phenomenon, given that the X-Zylo is an axisymmetric device, is that a torque is applied that changes the direction of the angular momentum vector. The definition of a torque is,

$$\Lambda = \vec{r} \times \vec{F} \tag{4.49}$$

Since the change experienced in the angular momentum manifests itself as a yaw (as opposed to pitch or roll), it follows that there exists an asymmetry in the forces between the upper and lower surfaces. To produce the yaw-left result, there must either be

1. A greater force in the direction of flight on the upper surface, or
2. A greater force in the direction opposite of flight on the lower surface

By considering the effect of two flat-plate vortex distributions it can be seen that the plate that is staggered behind the other will also have a reduced lift due to the influence of the front vortex element. Conversely, in a situation with positive stagger, the upper element receives an effective increase in angle of attack and therefore creates proportionally more lift. Schematics of these situations are seen in Figure 4.6.

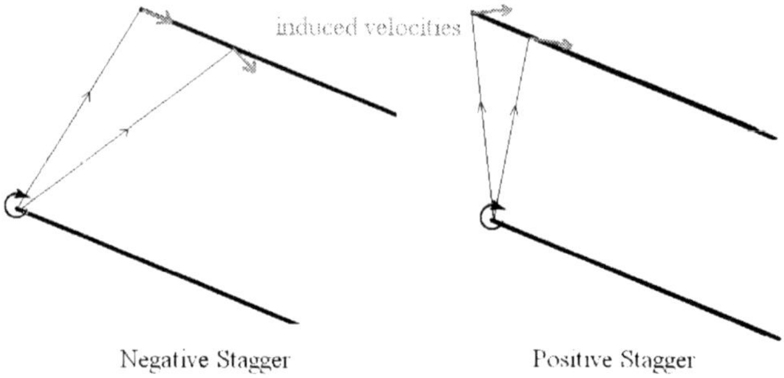

Figure 4.6. The relative influence of vortex elements from coupled vortex sheets.

This idea is confirmed and quantified by [7]. Although the geometric stagger of the X-Zylo is zero, the effective stagger is proportional to the angle of attack, α. As α increases, the relative influence of the leading vortices on the trailing vortices increases, skewing the lift distribution more in favor of the leading (lower) plate. Munk established that the change of the induced angle of attack for each plate is the same magnitude and can be found by:

$$\beta_1 = \frac{C_L^2}{2\pi^2 B}\frac{S}{b^2}\left(\frac{1}{k^2} - 0.5\right)\frac{G}{R} \qquad (4.50)$$

where G represents the gap between airfoils, S is the planform area, b is the span, B is an efficiency factor,

k is the elliptic modulus, and R is the average distance over which a vortex element on one airfoil exerts its influence on the other. Interestingly enough, the average value of R can be calculated using a combination of elliptic integrals. For the current example we take the values to be $G = d, S = \pi dc$, and $b = 2d$. Using these values along with the tables of the elliptic integral solutions provided by Munk, and substituting $CL = 2\pi\alpha$ (as we found in solving the lift for a flat plate) the induced change in angle of attack for our geometry is

$$\beta_1 = \pm 0.2051\,\alpha^2 \qquad (4.51)$$

As the angle of attack varies, so does the non-dimensional momentum thickness, θ^*, along the boundary of the airfoil. Although a higher angle of attack means lower wall shear, the momentum defect at the trailing edge more effectively measures the rate of momentum lost to viscous dissipation through the boundary layer as α becomes such that $\sin\alpha$ does not equal zero. The drag over the plate can be described as:

$$D_{\theta^*} = \rho\, u_\infty^2\, \theta^* \sqrt{\frac{u_\infty}{\nu\, c}} \qquad (4.52)$$

where the non-dimensional term, θ^*, is dimensionalized by the last term, and the profile of θ^* versus α is shown in figure 4.7.

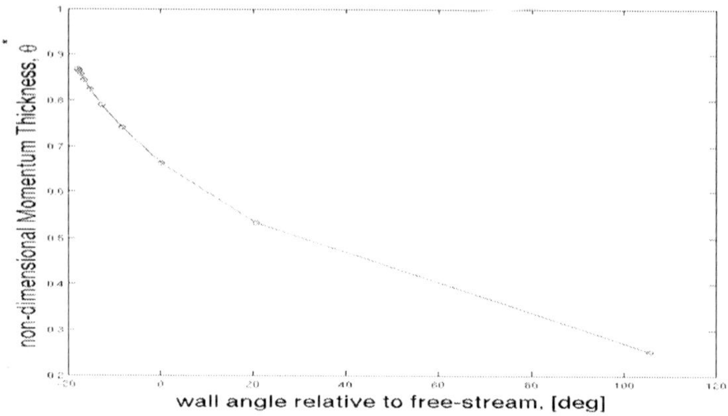

Figure 4.7. Relationship between flat-plate angle of attack and non-dimensional momentum thickness.

Since this effect only is significant when the flight velocity is low, the example calculation of the torque will not be for the standard launch conditions, but rather for the conditions of $u_{inf} = 5\ m/s$, $\omega = 15\ rad/s$ and $\alpha = 12.62$ degrees.

As the X-Zylo flies, viscosity acts to reduce both its rotation speed and its forward speed, increasing its susceptibility to forces attempting to alter its trajectory. These reduced quantities are estimations of typical flight values that will hopefully capture the essence of the behavior. Prior to the calculations there are several considerations that should be mentioned.

Firstly, the lift itself cannot produce a torque since we have assumed that pressures can only act normal to the plate, which is parallel to *r* such that $r \times L = 0$. Second, since the aerodynamic center aligns with the center of mass there is no net force applied that could produce a torque. Therefore the likely culprit is the uneven drag due to the biplane-like interference between lifting elements. With the value for α known, we can calculate Munk's induced angle of attack parameter to be $\beta_1 = \pm 0.57°$, and the corresponding angles of attack for the upper and lower surfaces to be:

$$\alpha_{upper} = 12.049° \quad \text{and} \quad \alpha_{lower} = 13.20° \tag{4.53}$$

and the corresponding drag forces due to momentum defects are:

$$D_{\theta^*,upper} = 1.6079 \times 10^{-3}\, N \quad \text{and} \quad D_{\theta^*,lower} = 1.6170 \times 10{-}3\, N \tag{4.54}$$

From this the net torque is calculated to be:

$$\Lambda = 4.4075 \times 10^{-7}\ N \cdot m \qquad (4.55)$$

Since we know the value of the angular momentum, $\Omega = 8.02 \times 10^{-4}\ kg\ m^2/s$, we can find the rate of gyroscopic precession as:

$$\frac{d\phi}{dt} = \frac{\Lambda}{\Omega}\frac{180}{\pi} = 0.03146\ deg/s \qquad (4.56)$$

Unfortunately this value falls well short of the observed behavior.

A colleague suggested that the effect of the relative positions of the center of mass and the center of lift should be revisited. In equation 4.28 it was shown that the theoretical center of pressure is 1/4 chord. Since the center of mass was measured at 0.238 chord the values were deemed suitably close to neglect their effects. Because no aerodynamically subtle explanation was validated for the yaw tendency, we turn back to this previously neglected relationship.

For the high α case, the same procedure as above is followed to determine the rate of yaw.

$$\frac{d\phi}{dt} = \frac{(0.02\,c) \times L}{\Omega}\frac{180}{\pi} = 17.4\ deg/s \qquad (4.57)$$

With a moment arm of $0.02c$ the small torque produces a rate of yaw that is very close to what we have seen in practice! In theory the behavior can be eliminated with higher tolerances during manufacture that exactly align the two centers. However, the thin airfoil approximation breaks down as α increases, resulting in boundary layer separation and other viscous effects that move the center of pressure rearward. Figure 4.8 shows actual wind tunnel data for a NACA-0015 airfoil (thin, symmetric, 2D) that charts the increase in the moment about the leading edge as the angle of attack is increased. Because of this phenomenon there is little that better manufacturing tolerances can do to alleviate this behavior at high α conditions.

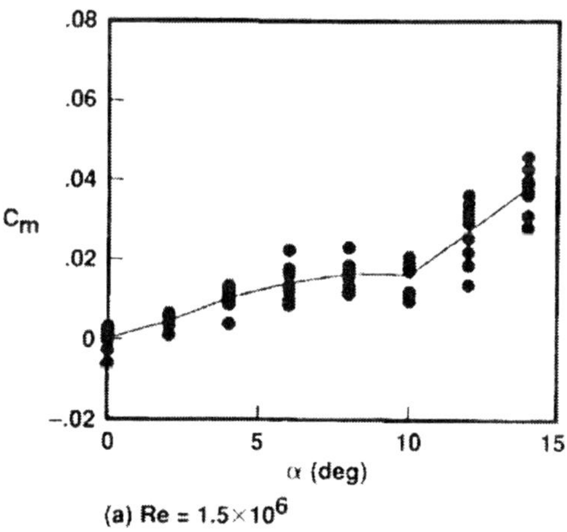

Figure 4.8. Change in moment coefficient about the leading edge with increasing α using NACA-0015 airfoil, [10]

It has been purported by some that the mechanism for this is the Magnus effect. This is not the case. As the X-Zylo slows and begins to fall, increasing the angle of attack, the force due to the Magnus effect does increase, but the maximum achievable angle of attack prior to stall is still such that the ratio of forward velocity to vertical velocity, $\tan\alpha \sim 1$. Therefore the dominating dynamic pressures still act along the cylinder's axis,

$$\frac{1}{2}\rho u_{\parallel}^2 \gg \frac{1}{2}\rho u_{\perp}^2$$

negating the relative effect of the Magnus force.

6 Experiments

Several qualitative experiments have been carried out to verify whether the theory outlined above can properly explain the aerodynamics of the X-Zylo and to answer several of the questions laid out in the introduction.

Altered Center of Mass – The hypothesis that the relative location of the center of mass and the center of pressure is the determining factor in the gyroscopic precession was tested on the actual X-Zylo. By sliding a uniform distribution of paper-clips onto the thin trailing edge, the center of mass was moved to 0.45 chord. As Figure 5.1 shows, if the center of mass is located aft of the center of pressure, the applied torque will be into the paper. We know from freshman physics that when a torque is applied to a rotating mass (a gyroscope) the new angular momentum vector is the vector sum of the torque vector and angular momentum vector. Therefore, for a clockwise spinning X-Zylo, a torque out of the board results in a left-hand yaw, and a torque into the board results in a right-hand yaw. If this hypothesis is true, the X-Zylo with the center of mass moved rearward should curve the opposite way of the unweighted X-Zylo.

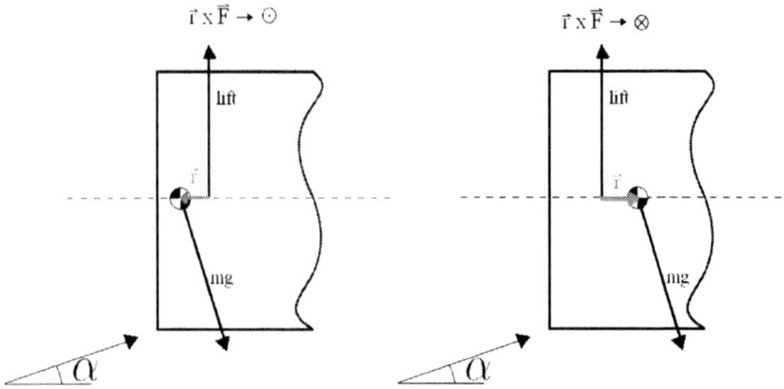

Figure 5.1. Comparison of torques produced by relative locations of the center of mass and center of pressure.

When the X-Zylo is thrown in the weighted configuration, the trajectory does curve to the right! As suggested, the gyroscopic precession is not a result of subtle aerodynamic phenomena, but simply of the weighting of the device. When the weighted X-Zylo is thrown, the flight is very unstable and prone to tumble, so large angular velocities are needed to observe the right-turning tendency. The reason for this instability is that having the weight in front of the center of pressure produces a stabilizing torque along the axis of the cylinder. Much like a penny is clipped to the nose of a balsa-wood airplane, having the bulk of the inertia in front of the location where pressure acts maintains stability. In the weighted configuration this relationship is reversed and the small drag force at the center of pressure acts to tumble the device out of stable flight.

Figure 5.2. The weighted X-Zylo with center of mass at 0.45c

Effect of Wavy Trailing Edge – It was hypothesized that the wavy trailing edge on the X-Zylo is purely for aesthetics. Since the induced drag is only a function of the lift produced, the details of the wake should not be important - they are merely a different tool by which the same information can be gathered. Accordingly, a simple glove was fitted to the exterior of the X-Zylo (Figure 5.3) such that the chord-length, thickness, and center of mass were not appreciably affected. It became apparent from flight tests that there is no deviation in the basic behavior: the X-Zylo flew equally well and still exhibited the trademark left-yaw of a precession-rate typical of gloveless flights. Upon conclusion of these tests all parties involved were satisfied that the wavy trailing edge indeed has no effect on overall performance.

Can it be Thrown Backwards? – With the glove still on (so as to not produce leading-edge vortices that would undoubtedly hinder stable flight) the X-Zylo was thrown backwards. Not surprisingly, it does not work. As we showed earlier, it is an important feature that the center of mass is slightly forward of the aerodynamic center. When they do not, destabilizing moments are produced that make stable flight very difficult.

Sudden Gust of Crosswind – To examine the precessional behavior in more detail a sharp corner between buildings was sought to provide an immediate crosswind shortly after stable flight was established. Upon finding a suitable location, the test was carried out with the expected results. When the X-Zylo is given counter clockwise rotation (as viewed from behind), the effect of a crosswind (from X-Zylo's left to right) of approximate speed 5 *m/s* is to pitch the X-Zylo upward. Since the effect of an increased angle of attack vertically is to yaw the X-Zylo left, the fact that a crosswind produces a pitch is consistent with experience. If we measure the torques from the center of mass, we see that a lateral force on the center of pressure produces a torque in the vertical direction, pitching the vector Ω upward. As we would expect from a gyroscope, an applied force manifests itself 90 degrees from where it acts.

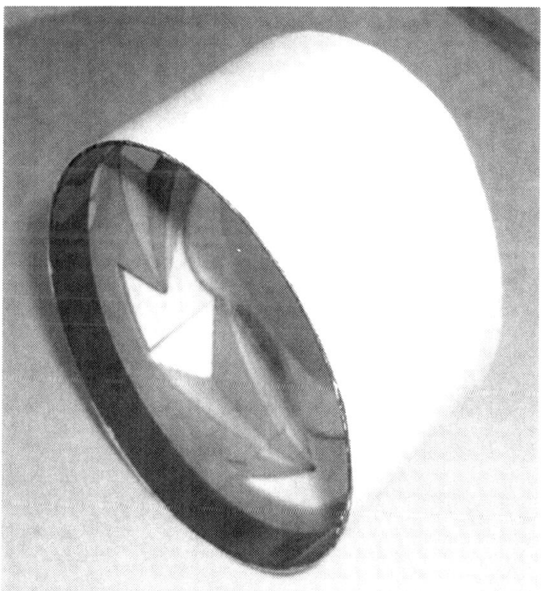

Figure 5.3. X-Zylo with fitted glove attached

7 Conclusions

Through the analysis conducted, we have shown that the lift mechanism of the X-Zylo is most likely akin to the lift generated by an infinite, two-dimensional flat plate. The analytical result for the angle of attack at standard launch conditions is consistent with the experimentally observed behavior. As well, the viscous effects on rotation were quantified to a value reasonably consistent with experiment using correlations generated from the Falkner-Skan set of thin shear layer equations. Lastly attempts were made to explain the precession of the gyroscope at low forward velocity and low angular momentum conditions as is present toward the end of typical flights. The mechanism by which this precession occurs has been explained and quantified with the basic theories of physics and flight, and experimentally proved on the actual device.

The results of this investigation suggest that a time-accurate simulation of the X-Zylo's flight could be created using the methods outlined here.

References

[1] The Gyroball. Sports Center, ESPN, 2007.

[2] Aerodynamics for Engineers. Prentice Hall, 2002.

[3] Fluid Mechanics. Elsevier Academic Press, 2004.

[4] Mark Drela. Class notes from 16.13 Aerodynamics of Viscous Fluids.

[5] Luciano Demasi. Investigation on conditions of minimum induced drag of closed wing systems and c-wings. Journal of Aircraft, 2007.

[6] Engineering Fluid Mechanics. John Wiley and Sons, Inc., 2001.

[7] M.M. Munk. General biplane theory. Technical report, National Advisory Committee for Aeronautics, 1923.

[8] M.M. Munk. The air forces on a systematic series of biplane and triplane cellule models. Technical report, Langley Memorial Aeronautical Laboratory, 1927.

[9] C.B. Millikan. An extended theory of thin airfoils and its application to the biplane problem. Technical report, California Institute of Technology, 1931.

[10] K.W. McAlister and R.K. Takahashi. Naca 0015 wing pressure and trailing vortex measurements. Technical report, US Army Aviation Systems Command, 1991.

About the Author

David L. Tarr is a naval aviator who became interested in flying as a small boy. Growing up in Seattle, he went to airshows and watched the flying machines including the Navy Blue Angels, all of which were great influences on his career choice to pilot the F/A18E Super Hornet.

In addition to piloting airplanes, he has always had fun with toys that fly, the X-Zylo being among his favorites.

David earned his B.S. in Mechanical Engineering from the University of Colorado and his S.M. in Aeronautics and Astronautics from the Massachusetts Institute of Technology. It was the fertile environment of M.I.T. that led to the graduate student debate about the X-Zylo's flight characteristics.

The X-Zylo™ can be purchased at stores throughout the world. Buy one and test your own theories.

CPSIA information can be obtained at www.ICGtesting.com
Printed in the USA
LVOW11s1449270416

485579LV00001B/104/P

9 780982 114834